IGNITION:

THE OVERVIEW OF JET ENGINE DESIGN

BY

BENTO GYPSON

TABLE OF CONTENT

INTRODUCTION

Jet Engine Overview

A jet engine is a type of propulsion system that accelerates a stream of air to generate thrust, which propels an aircraft forward. There are several types of jet engines, including turbojet, turbofan, turboprop, and turbo shaft engines, each designed for specific applications.

Components of a Jet Engine:

Inlet: The engine's inlet allows air to enter the engine. In some jet engines, the inlet also includes devices to regulate the airflow and prevent excessive speed during high-velocity flight.

Compressor: The compressor consists of rotating and stationary blades that compress incoming air. Compressing the air increases its pressure and energy, preparing it for combustion.

Combustion Chamber: In the combustion chamber, fuel is injected into the compressed air and ignited. This mixture burns at high temperatures, creating a high-speed exhaust gas.

Turbine: The turbine is a set of blades connected to a shaft. As the high-temperature, high-speed exhaust gases flow over the turbine blades, they cause the turbine to spin.

Nozzle: The nozzle is responsible for accelerating the exhaust gases to produce thrust. By expelling the high-speed gases backward, the engine generates forward thrust according to Newton's third law of motion.

CHAPTER ONE

History of jet engine

The history of the jet engine spans several centuries, with key developments and innovations leading to the powerful and efficient engines we have today. Here's an overview of the significant milestones in the history of jet engines:

Early Concepts (17th-18th centuries): The concept of jet propulsion can be traced back to the 17th century when Sir Isaac Newton described his third law of motion, which states that for every action, there is an equal and opposite reaction. However, practical applications were not developed until centuries later.

Hero's Engine (1st century AD): Although not a jet engine in the modern sense, Hero's engine, also known as an aeolipile, was a simple steam-powered device invented by Hero of Alexandria in the 1st century AD. It demonstrated the basic principle of jet propulsion.

Pioneering Efforts (Early 20th century): Early pioneers like Sir Frank Whittle in England and Hans von Ohain in Germany independently worked on the concept of jet propulsion in the early 20th century. Whittle's work led to the first practical jet engine designs.

First Jet Engine (1930s-1940s): Frank Whittle, a British engineer, and Hans von Ohain, a German engineer, both developed functional jet engines in the 1930s. Whittle's engine, the W.1, powered the Gloster E.28/39, which made its first successful flight in 1941. In Germany, the Messerschmitt Me 262 became the first operational jet-powered fighter aircraft during World War II.

Post-War Developments (1940s-1950s): After World War II, jet engine technology rapidly advanced. Gas turbine engines were developed for commercial aviation, leading to the introduction of the first jet airliners like the de Havilland Comet and the Boeing 707 in the 1950s.

Turbofan Engines (1950s-1960s): The development of turbofan engines in the 1950s significantly improved the efficiency and noise levels of jet engines. Turbofan engines became the standard for most commercial airliners, offering better fuel efficiency and reduced noise pollution.

Supersonic Flight (1950s-1960s): The 1950s and 1960s saw the development of supersonic jet aircraft, including the iconic

Concorde, jointly developed by British and French engineers. The Concorde entered commercial service in 1976 and remained in operation until 2003.

Modern Advancements (Late 20th century-Present): Continued research and advancements in materials, aerodynamics, and computer technology have led to the development of highly efficient and powerful jet engines used in modern commercial and military aircraft. These engines are more fuel-efficient, quieter, and environmentally friendly compared to earlier designs.

The history of jet engines reflects the continuous innovation and collaboration of engineers and scientists over centuries, leading to the sophisticated propulsion systems we rely on for air travel and various other applications today.

CHAPTER TWO

Types of Jet Engines

Turbojet Engines: Turbojets are the simplest type of jet engine. It consist of a compressor, combustion chamber, turbine, and nozzle. Turbojets are used in high-speed military aircraft and some early commercial planes.

Turbofan Engines: Turbofan engines are the most common type of engine in commercial aviation. They have an additional fan at the front that bypasses a portion of the incoming air around the engine core. This bypass air provides more additional thrust and which improves fuel efficiency.

Turboprop Engines: Turboprop engines power propellers, which are used in regional and smaller aircraft. These engines combine a jet engine core with a gearbox to drive the propeller.

Turboshaft Engines: Turboshaft engines are used in helicopters and other rotorcraft. They provide power to the rotor through a shaft and are also used in various industrial applications.

Jet engines have revolutionized aviation and are essential for modern air travel, enabling aircraft to travel faster, farther, and more efficiently.

The fluid used for jet to operate

Jet engines use various fluids for their operation, each serving a specific purpose:

Jet Fuel: The primary fluid used in jet engines is jet fuel, which is a specific type of aviation fuel. Jet fuel powers the combustion process in the engine, creating the high-temperature, high-speed exhaust gases necessary for generating thrust. The most common types of jet fuel are Jet A and Jet A-1, which are kerosene-based fuels.

Oil: Jet engines require oil for lubrication and cooling. Oil is used to lubricate various engine components, such as bearings and gears, to ensure smooth operation. It also helps in dissipating heat generated during engine operation.

Hydraulic Fluid: Jet engines often incorporate hydraulic systems for various functions, such as controlling aircraft landing gear, flaps, and other movable surfaces. Hydraulic fluid is used in these systems to transmit power and operate hydraulic actuators.

These fluids are critical for the safe and efficient operation of jet engines and the aircraft they power. Proper maintenance and monitoring of these fluids are essential to ensure the engine's reliability and performance.

CHAPTER THREE

How jet engine operate

Jet engines operate through the principle of Newton's third law of motion, which states that for every action, there is an equal and opposite reaction. Here's a step-by-step explanation of how jet engines work:

Intake: Air is drawn into the jet engine through the intake. The intake is designed to compress the incoming air and regulate its speed, especially during high-velocity flight.

Compression: The incoming air passes through a series of rotating and stationary blades called the compressor. The compressor compresses the air, increasing its pressure and energy.

Combustion: Fuel is injected into the highly compressed air in the combustion chamber. The mixture of fuel and air is then ignited. The burning process releases a tremendous amount of heat, which causes the air to expand rapidly, creating high-speed exhaust gases.

Turbine: The high-speed exhaust gases flow over a set of turbine blades, causing them to spin. This spinning motion drives the compressor, maintaining the airflow and compression process.

Nozzle: The now high-speed exhaust gases exit the engine through a nozzle at the back. As the gases are expelled backward, they create a forward thrust on the engine, according to Newton's third law. This thrust propels the aircraft forward.

Thrust and Propulsion: The forward thrust generated by the jet engine propels the aircraft through the air. By adjusting the engine's thrust, pilots can control the speed and direction of the aircraft.

This continuous process of intake, compression, combustion, and exhaust creates a powerful stream of high-speed exhaust gases, which provide the necessary thrust to propel the aircraft forward. Jet engines are incredibly efficient at converting fuel into thrust, making them the primary propulsion system for most modern aircraft.

CHAPTER FOUR

Jet Engine Design Evolution

Design of Jet Engines: Unraveling the Power of Flight

Jet engines, marvels of engineering ingenuity, are the beating hearts of modern aviation. Their design represents a delicate balance of precision, power, and efficiency, enabling aircraft to soar across the skies at incredible speeds. Let's delve into the intricacies of their design, exploring the fundamental principles and cutting-edge technologies that make them the driving force behind air travel. The design of jet engines represents a remarkable fusion of scientific principles, advanced materials, and creative engineering solutions. As aviation continues to push the boundaries of what is possible, the evolution of jet engines remains at the forefront, propelling humanity towards a future where the skies are no longer limits but pathways to new horizons.

Basic Components:

At its core, a jet engine comprises several key components, including the compressor, combustion chamber, turbine, and nozzle. The process begins as the compressor draws in vast quantities of air, compressing it to increase its pressure. This high-pressure air then enters the combustion chamber where it mixes with fuel, creating a combustible mixture. Igniting this mixture produces a high-speed exhaust jet, propelling the aircraft forward.

Operating Principles:

Jet engines operate based on Sir Isaac Newton's third law of motion: for every action, there is an equal and opposite reaction. By expelling a fast-moving stream of gas backward, the engine generates a forward thrust, pushing the aircraft through the air. This principle underpins the design of both commercial airliners and military fighter jets, showcasing the versatility of jet engine technology.

Turbojet vs. Turbofan:

Turbojet engines, the earliest form of jet propulsion, have evolved into more efficient variants known as turbofan engines. Turbojets are characterized by their simple design and high-speed capabilities, making them ideal for supersonic aircraft. On the other hand, turbofan engines incorporate a large fan at the front, which accelerates a portion of the incoming air around the engine, enhancing efficiency and fuel economy. These engines are commonly found on commercial airplanes, providing the thrust necessary for long-haul flights.

Advancements in Materials and Design

The materials used in jet engine components have seen significant advancements over the years. High-strength alloys and composite materials are now commonplace, ensuring the engine's durability under extreme conditions. Computational simulations and aerodynamic optimizations have also played a pivotal role in refining the design, enhancing performance, and reducing environmental impact.

Environmental Considerations:

With a growing emphasis on environmental sustainability, jet engine design has shifted towards minimizing carbon emissions and noise pollution. Researchers and engineers are exploring innovative solutions, such as quieter engine designs and alternative fuels, to create a more eco-friendly aviation industry.

Future Innovations:

The future of jet engine design holds exciting prospects, including the development of hypersonic engines capable of reaching speeds beyond Mach 5. Additionally, electric propulsion and hybrid technologies are being explored to revolutionize short-haul flights, paving the way for a greener and more efficient aviation sector.

Mathematical design of jet engine

The mathematical design of jet engines involves complex equations and principles from various branches of physics and engineering. While I can't provide a detailed mathematical derivation due to the limitations of this format, I can outline the fundamental concepts and equations involved in the mathematical design of jet engines.

Thermodynamics:

Jet engines operate based on the principles of thermodynamics. The Brayton cycle, a thermodynamic cycle that describes the functioning of a jet engine, involves four main processes: isentropic compression, constant pressure combustion, isentropic expansion, and constant pressure exhaust. The efficiency of the Brayton cycle can be calculated using the following equation:

$$Efficiency = 1 - (\frac{T_2}{T_1})^{\gamma-1}$$

Where T_1 and T_2 are the temperatures at the inlet and outlet of the engine, and γ (gamma) represents the ratio of specific heats.

Thrust Equation:

The thrust produced by a jet engine can be calculated using the following equation, derived from Newton's second law of motion:

$$F = \dot{m} \times (V_e - V_0) + (p_e - p_0) \times A_e$$

Where F is the thrust, $\dot{m}$ is the mass flow rate of the exhaust gases, V_e is the exhaust velocity, V_0 is the aircraft velocity, P_e is the exhaust pressure, P_0 is the ambient pressure, and A_e is the exit area of the engine.

Compressor and Turbine Equations:

The compressor and turbine in a jet engine are essential components responsible for air compression and power generation, respectively. The work done by the compressor and turbine can be calculated using the following equations:

$$Work\ Compressor = C_P \times T_0 \times \left(\left(\frac{P_2}{P_1}\right)^{\frac{\gamma-1}{\gamma}} - 1\right)$$

$$Work\ Turbine = C_P \times T_0 \times \left(1 - \left(\frac{P_4}{P_3}\right)^{\frac{\gamma-1}{\gamma}}\right)$$

Where C_P is the specific heat at constant pressure, T_0 is the total temperature, P_1, P_2, P_3 and P_4 represent pressures at various points in the engine.

Specific Fuel Consumption (SFC):

SFC is a crucial parameter that measures the efficiency of an engine. It represents the amount of fuel required to produce a unit of thrust for a specific duration. SFC can be calculated using the following equation:

$$SFC = \frac{\dot{m}_{fuel}}{F}$$

Where $\dot{m}_{fuel}$ is the mass flow rate of the fuel.

These equations provide a basic overview of the mathematical principles involved in the design and analysis of jet engines. Actual jet engine design involves more intricate calculations, including considerations for various components, thermodynamic efficiencies, and real-world factors that affect performance and safety. Engineers use specialized software and computational tools to perform detailed simulations and optimizations for practical jet engine designs.

CHAPTER FIVE

Effect of Nanotechnology in the design of jet engine

Nanotechnology, the manipulation of materials at the nanoscale level, has revolutionized various industries, including aerospace engineering. Nanotechnology plays a pivotal role in enhancing various aspects of jet engine design. By leveraging nanomaterials and innovative manufacturing techniques, engineers can create lighter, stronger, and more efficient engines. These advancements not only improve the performance of aircraft but also contribute to reducing environmental impact by enhancing fuel efficiency and minimizing emissions. In the context of jet engine design, nanotechnology contributes significantly in several ways:

1. Lightweight and Strong Materials:

Nanotechnology enables the development of advanced composite materials at the molecular level. These materials are incredibly lightweight and possess exceptional strength and durability. Integrating nanocomposites into the design of jet engine components reduces overall weight, enhancing fuel efficiency and increasing the engine's performance.

2. Enhanced Engine Coatings:

Nanomaterials are utilized to create specialized coatings for jet engine components. These coatings provide increased resistance to wear, corrosion, and high temperatures. By protecting crucial parts such as turbine blades and combustion chambers, these coatings prolong the engine's lifespan and improve its reliability, reducing maintenance requirements and operational costs.

3. Improved Heat Management:

Nanotechnology allows for the development of thermal barrier coatings (TBCs) with nano-sized particles. TBCs are applied to engine components exposed to extreme heat, such as turbine blades. These coatings provide excellent thermal insulation, preventing heat from damaging the underlying metal. By improving heat management, jet engines can operate at higher temperatures, leading to increased efficiency and power output.

4. Nanofluids for Cooling:

Nanofluids, engineered by dispersing nanoparticles in coolants, are used in jet engines for efficient cooling. These nanofluids have superior thermal conductivity, ensuring more effective heat dissipation from critical engine components. By optimizing the cooling process, the engine's overall efficiency and performance are enhanced.

5. Fuel Efficiency and Emissions Reduction:

Nanotechnology contributes to the development of catalysts and filters made of nanostructured materials. These catalysts help in refining aviation fuels and reducing harmful emissions, making jet engines more environmentally friendly. Additionally, nanomaterial-based sensors are used to monitor fuel combustion in real-time, allowing for precise control and optimization of the combustion process, thereby improving fuel efficiency.

6. Nanoscale Manufacturing:

Nanotechnology enables precise control over manufacturing processes at the atomic and molecular levels. Techniques like nanoimprinting and molecular self-assembly facilitate the production of intricate engine components with high precision. This level of precision ensures the optimal performance of the engine, leading to improved efficiency and reliability.

CHAPTER SIX

The use of jet engine

Jet engines are used in various applications, primarily in the field of aviation, but also in other industries. Overall, jet engines play a crucial role in transportation, defense, space exploration, and power generation, enabling faster travel, efficient energy production, and advancements in various fields of technology and industry. Here are the main uses of jet engines:

Aircraft Propulsion: Jet engines are the primary means of propulsion for most modern airplanes and helicopters. They power commercial airliners, military aircraft, private jets, and other types of planes, allowing for fast and efficient air travel.

Military Applications: Jet engines are widely used in military aircraft, including fighter jets, bombers, reconnaissance planes, and unmanned aerial vehicles (drones). Military jets are designed for high speed, agility, and maneuverability, making them essential for national defense and military operations.

Space Exploration: Rocket engines, a specialized type of jet engine, are used in space exploration. They provide the necessary thrust to launch spacecraft into space and propel them during various stages of their missions. Jet engines are also used in spaceplanes for atmospheric flight.

Marine Propulsion: Jet engines, particularly water jet propulsion systems, are used in some high-speed boats and ships. Water jets provide greater maneuverability and speed compared to traditional propellers, making them suitable for military vessels, ferries, and luxury yachts.

Power Generation: Jet engines, especially gas turbine engines, are used in power generation plants to produce electricity. These engines are connected to generators, converting the rotational energy from the engine into electrical power. Gas turbine power plants are efficient and can be used for peaking power and emergency backup.

Ground Transportation: Although less common, jet engines have been used experimentally in ground vehicles, such as cars and buses. These applications are often prototypes or concepts aimed at exploring alternative propulsion methods for high-speed ground transportation.

Industrial Applications: Jet engines, particularly gas turbines, are used in various industrial settings to drive machinery, pumps, and compressors. They are favored for their high power-to-weight ratio and reliability in continuous operation.

CHAPTER SEVEN

Advanced jet engine

Advanced jet engines, often referred to as next-generation or advanced propulsion systems, are at the forefront of aerospace engineering. These engines incorporate cutting-edge technologies to improve efficiency, reduce emissions, enhance safety, and enable higher performance. These advancements represent the ongoing efforts in the aerospace industry to create more efficient, eco-friendly, and high-performance jet engines, shaping the future of aviation and propulsion technologies. Here are some key features and advancements in advanced jet engines:

High-Bypass Turbofan Design: Modern advanced jet engines often feature high-bypass turbofan designs. These engines have a larger fan at the front, allowing a significant portion of the incoming air to bypass the combustion process. This bypass air enhances efficiency and reduces fuel consumption.

Improved Materials: Advanced jet engines utilize lightweight and high-strength materials, such as advanced alloys and composites. These materials improve engine durability, reduce weight, and enhance overall efficiency.

Efficiency Enhancements: Advanced engines incorporate technologies like advanced aerodynamics, optimized turbine designs, and improved thermal management. These enhancements increase the overall efficiency of the engine, translating to reduced fuel consumption and lower operating costs.

Noise Reduction: Advanced noise reduction technologies, such as chevrons on engine nozzles and redesigned fan blades, help mitigate noise pollution during takeoff and landing. Quieter engines contribute to more comfortable airport environments and reduced impact on surrounding communities.

Emission Reduction: Advanced jet engines feature designs to reduce harmful emissions. Combustion technologies, such as lean-burn and staged combustion, help minimize nitrogen oxide (NOx) emissions. Additionally, research focuses on alternative fuels, including biofuels and synthetic fuels, which have the potential to lower greenhouse gas emissions.

Digitalization and Predictive Maintenance: Advanced engines are equipped with sensors and data analysis systems that enable real-time monitoring of engine performance. Predictive maintenance algorithms analyze data to predict potential issues, allowing airlines to perform maintenance tasks proactively, reducing downtime and increasing operational efficiency.

Adaptive Cycle Engines: Research is ongoing in the development of adaptive cycle engines. These engines can adjust their configuration during flight, optimizing performance based on specific flight conditions. Adaptive cycle engines promise greater fuel efficiency and versatility across a wide range of operating conditions.

Hypersonic Propulsion: Advanced jet engines are crucial in the development of hypersonic flight technologies. These engines are designed to propel aircraft and missiles at speeds exceeding Mach 5 (hypersonic speeds). Hypersonic propulsion presents challenges related to extreme temperatures and pressures, demanding innovative materials and cooling techniques.

Electrification and Hybridization: Research is exploring the integration of electric and hybrid propulsion systems with jet engines. These hybrid systems can enhance efficiency, especially during low-power and standby phases of flight. Electric propulsion is also being investigated for smaller aircraft, contributing to the development of urban air mobility solutions.

CHAPTER EIGHT

Benefits of jet engine

Jet engines offer several benefits, which contribute to their widespread use in various applications:

High Speed and Efficiency: Jet engines allow aircraft and other vehicles to travel at high speeds, making air travel and military operations faster and more efficient. They have a high power-to-weight ratio, enabling rapid acceleration and deceleration.

Long Range: Jet engines provide the necessary thrust for long-distance travel, allowing airplanes to cover vast distances without frequent refueling. This long-range capability is essential for commercial aviation, connecting cities and countries across the globe.

Altitude Capability: Jet engines are capable of operating at high altitudes, making them suitable for both commercial and military aircraft. This ability is crucial for flying over mountainous terrain and reaching the cruising altitudes necessary for long-distance flights.

Versatility: Jet engines are versatile and can be adapted for various applications, including commercial aviation, military aircraft, helicopters, ships, power plants, and space exploration. This versatility makes them valuable in different industries.

Improved Safety: Modern jet engines are equipped with advanced safety features, such as redundant systems and computerized controls, making air travel safer for passengers. Continuous advancements in engine technology enhance reliability and reduce the risk of failures.

Reduced Travel Time: Jet engines significantly reduce travel time, enabling people and goods to move quickly between distant locations. This benefit is particularly important in emergency situations, where rapid response and transportation are essential.

Economic Impact: Jet engines support a vast industry, including aircraft manufacturing, maintenance, airlines, and related services. This industry generates significant economic activity, providing jobs and contributing to the global economy.

Environmental Efficiency: While jet engines emit greenhouse gases and other pollutants, advancements in engine technology have led to more fuel-efficient and environmentally friendly designs. Ongoing research aims to further reduce emissions and improve fuel efficiency.

Innovation and Research: The development of jet engines has driven innovations in materials, aerodynamics, and manufacturing processes. Research in jet engine technology continues to lead to improvements in efficiency, safety, and environmental impact.

Overall, the benefits of jet engines have revolutionized transportation, defense, and various industries, making the world more connected and accessible while driving technological advancements and economic growth.

Disadvantages of jet engine

While jet engines offer numerous advantages, they also come with several disadvantages and challenges:

Environmental Impact: Jet engines emit greenhouse gases, nitrogen oxides, and particulate matter, contributing to air pollution and climate change. Aircraft emissions can have a significant impact on the environment, especially at high altitudes.

Noise Pollution: Jet engines produce higher noise pollution, especially during takeoff and landing. Aircraft noise can disturb communities near airports, affecting the quality of life for residents and wildlife.

Fuel Consumption: Jet engines consume vast amounts of fuel, making air travel a major contributor to global oil consumption. This reliance on fossil fuels raises concerns about energy security and environmental sustainability.

High Initial Costs: Aircraft equipped with jet engines are expensive to manufacture and maintain. The initial investment for purchasing aircraft and building infrastructure, such as airports, is substantial.

Maintenance and Repairs: Jet engines require regular maintenance and periodic overhauls to ensure safe and efficient operation. Maintenance tasks are complex and require skilled technicians, leading to high maintenance costs.

Limited Altitude and Speed: While jet engines operate efficiently at high altitudes, they have limitations in extreme conditions. For example, they are not suitable for space travel due to the absence of air in space. Additionally, jet engines have speed limitations, which are approached by supersonic aircraft.

Dependency on Runways: Jet-powered aircraft require long runways for takeoff and landing. This dependency restricts their usability in areas with limited runway infrastructure, such as remote regions and densely populated urban areas.

Safety Concerns: Although modern jet engines are highly reliable, accidents and failures can still occur, leading to catastrophic consequences. Engine malfunctions, bird strikes, and other factors can pose safety risks.

Resource Intensity: The manufacturing and maintenance of jet engines require significant resources, including metals, materials, and energy. This resource intensity raises concerns about sustainability and resource depletion.

Impact on Wildlife: Aircraft, especially in busy airports, pose a threat to bird populations. Bird strikes can damage jet engines, leading to safety risks for both passengers and wildlife.

Addressing these disadvantages requires ongoing research and development to improve engine efficiency, reduce emissions, mitigate noise pollution, and explore alternative propulsion technologies.

CHAPTER NINE

Conclusion on jet engine

In conclusion, the jet engine stands as one of the most transformative inventions in the history of transportation and aviation. From its early conceptualization to the modern, highly efficient engines used in commercial airliners and military aircraft, the jet engine has revolutionized the way we travel, connect with one another, and explore the skies. Its ability to harness the principles of physics, particularly Newton's third law of motion, has propelled humanity into the jet age, enabling us to cross continents and oceans in a matter of hours, reach unprecedented speeds, and even venture into space.

However, the jet engine is not without its challenges. Environmental concerns, such as emissions and noise pollution, as well as the reliance on fossil fuels, have sparked ongoing research and innovation to create more sustainable and eco-friendly propulsion systems. Despite these challenges, the jet engine continues to evolve, with advancements in technology leading to quieter, more fuel-efficient engines that power the future of aviation.

In essence, the jet engine represents human ingenuity and the relentless pursuit of progress. It has reshaped the world, making it smaller, more connected, and accessible. As we move forward, the legacy of the jet engine will continue to inspire further advancements in aerospace engineering, ensuring that the skies remain a frontier of innovation and discovery.